AF725452

The Measure of All Things

The Measure of All Things

On the Relationship between Photography and Objects

Edited by Peter Weiermair

EDITION STEMMLE

Zurich – New York

William Henry Fox Talbot
Leaf, 1839–1840
Private collection

Contents

The Object as Touchstone for the Contemporary Experience of Reality and Aesthetic Strategies in Recent Photography

In the mid-19th century, during the infancy of photography as a new medium for the representation of reality, the motionless object was the primary focus of photographic depiction. Apart from the iconographic categories inherited from the traditional visual arts – portrait, nude and landscape – photography's predominant concern was the compilation of a visual record of the unlimited abundance of the world of objects. While the "still lifes" found among the early pictures testify to photography's inherent capacity to fix things permanently in arrested time (the vanity theme), the symbolic and allegorical levels of meaning that have historically characterized the "still life" painting now gave way to the positivistic, materialistic 19th-century concern with the depiction of reality. From the very outset, photography boasted the most immediate connection with the real. Despite the growing importance of virtual worlds, which are represented in our selection by Martin Dörbaum's fascinating contribution, photography remains dependent upon reality. Quite unlike other aesthetic languages – painting or literature, for instance – photography must, regardless of its power to alter and estrange, rely upon the external reality of the world of objects as the point of departure for such interventions. Of course, new scientific knowledge alone has changed the prevailing view of that external world in the 20th century, and the image of that world has been altered in photography as well. In view of the increasingly pervasive influence of processes involving the abstraction and estrangement of the real, concern with the immediate sensual presence of things has taken on an essential function. It seems a paradox that in a world in which we seem to lose sight of reality in the veritable flood of images, reality (which is its source) is not only to be revealed but made accessible to us once again through the very same medium of the photographic image. Much more than the still life, which has no importance as a concept in our context, the "measure of all things" itself plays a role of central significance. For the concept of measure points both to the process of change through the medium and to a view of the importance of things that is grounded in individual personal experience.

The seven positions featured here document the wealth of aesthetic strategies available to photographic artists today. They represent a clear rejection of the naive documentarism of the 19th century, to which they allude at most with a certain affectionate irony (Peter Peryer). Of importance are film, advertising and fashion contexts, aspects of scientific and technical photography and literature. Objects in the works of these seven artists are incorporated into varying strategies in which sequencing and serialization also play a major role. Major themes include not only differences in the way objects are viewed – differing photographic strategies and vocabularies, in other words – but also the artist's self-reflection on the categorical parameters of the medium itself. The anthology spans the spectrum from Muller's detached "objective" photos to the sophisticated works of Ann Mandelbaum, which occupy an in-between world of ambivalent meanings and functions of objects; it includes the pensive "memory images," focused upon individual object experiences, of the New Zealand photographer Peter Peryer as well as Vik Muniz's "translations," as illustrations of theories on semiotics and reception – a tight-rope walk between the artificial and the natural. And it also encompasses both the claustrophobic spatial projections of Alessandra Tesi and Tamara Grčić's picturesque-conceptual observations of situations in New York's Chinatown. All of these artists are concerned with the presentation of a perceptual process which places a world of everyday objects at the center of focus; and they are concerned with visual experiences in the medium of photography itself. Things are observed, registered, staged, documented, estranged from their usual context; they are actors upon the stage of everyday reality in a medium whose psychological and philosophical dimensions are taking on increasing importance.

Vik Muniz, a native of Brazil who now lives in New York, makes wire objects based on memories of images seen in photographs, and he then photographs them again. Muniz is interested in the differences in the media themselves and in the manner in which they replace reality and become signs – in the boundary between the sign and the concrete object. And he is also concerned with the processes of memory and their subtle, paradoxical mechanics. The interrelationships that link object, reproduction and the authentic photographic image of the sign he creates permit him to explore the different languages used to represent things and to capture the process of reception in paradoxical photographs. Much in the same way that Muniz invites the viewer to contemplate the essence of things and the phenomenon of memory, Martin Dörbaum uses virtual images – obviously of his own live-in studio – to construct a world of objects in which abstract and real elements are joined in a suggestive and unresolvably puzzling manner. His pictures are based upon reality and contain traces of that reality but are not painting. The self-imposed task of reading these photos and figuring out what is false and what is true – or real, in other words – and of determining what is natural and what is artificial involves the viewer in a precarious balancing act.

The American photographer Ann Mandelbaum removes everyday objects from their natural context by altering their function or showing them in extreme close-up and thus transports them into a richly associative in-between realm of potential erotic and sexual meanings.

Heinz Liesbrock has remarked that it is not what her photographs show, as images from the objective world, that gives them their powerful effect but rather the intensity of an immediate visual presence. This criterion can be applied to most of the contributions featured here – and especially to the work of Christopher Muller, whose photography could hardly contrast more vividly with that of Ann Mandelbaum. Where Mandelbaum concentrates on an intimate format and highly differentiated gray tones, Muller's photos are often quite large and intensely colorful. His pictures, which we might justifiably assign to the category of the contemporary still life, are painstakingly prepared in drawings. They represent a process of research into the fundamental question of how an image is created, an image that is entirely in accord with its own autonomous visual logic and thus takes its entirely artificial two-dimensionality seriously but at the same time reflects both the autonomy and the reality of the depicted object in the staged photograph. Muller's objects from everyday reality could easily have come from a department store catalogue, if it weren't for the dialogue Muller stages among them. He creates group portraits of often totally incompatible functional objects; these things are released from their functionality and they call to mind fairy tales in which inanimate objects begin to speak to each other when night falls. According to his own statements, he seeks the effect of surprise, the things' own reference to themselves. Tamara Grčić is also interested in combinations of things. Day after day she fixes them in her frame of view, in varying combinations on artificial grass somewhere in Chinatown. The photograph captures an everyday situation; the protagonists of daily life cause us to stop and think about the histories of these objects. Grčić is also concerned with ambiguity and multiple meaning in the photographic image. As a two-dimensional picture, the print represents a three-dimensional reality; the abstract and concrete qualities of these materials and objects are of equal importance.

Like Tamara Grčić, Alessandra Tesi also works with a wide variety of image-producing media, and this, it seems to me, is an essential characteristic of recent art. Tesi, an Italian artist who lives in Paris, has done large, intensely-colored prints that hang loosely from the wall, video installations and projections in which she prepares the wall in such a way that it appears as a shimmering skin that absorbs the image. Her investigation into the reality of used, no longer useful objects and time-worn hotel rooms – of the everyday objects that constantly surround us, has an existentialist undertone; her insistent observation has much of the intensity in the depiction of the world of objects that we find in the French *nouveau roman*.

Peter Peryer, a native of Australia, turned to photography quite late, and thus far he has deliberately kept his photographic oeuvre quite small, as if he wanted to live with only a few, milestone "memory images." His black-and-white photos spell out a world whose image calls to mind – albeit only at first glance – 19th-century pictures and the aesthetics of photography that prevailed during the first half of the 19th century. Examined more closely, however, they trigger a sense of productive irritation. These photos call upon the viewer to reject the unambiguous view, to think about the nature of the objects depicted and to reflect upon the illusions that originate in the aesthetics of photography. In Peryer's work – and the same can be said of the other artists represented here – the aspects of irritation in the experience of the pictures, of verification of reality and of testing for "truth" are of central and crucial significance.

Peter Weiermair
Exhibition Curator

Martin Dörbaum

Album/Detail

Der Wasserhahn ist zugesperrt. Ein Rest Wasser befindet sich noch auf dem Boden des Waschbeckens, da dieser eben ist. Es ist bereits zu wenig, um eine kontinuierliche Linie zu bilden und somit den Ablauf erreichen zu können. Daher zieht sich das Wasser aufgrund seiner ihm eigenen Oberflächenspannung in flachen blasenförmigen Gebilden zusammen. So wird es in den folgenden Stunden verdunsten. In der Wässerungsschale, die in nächster Nähe zum Wasserbecken steht und mit einem Zu- und Ablauf mit diesem verbunden ist, befindet sich noch Wasser. Es durchlief demnach die Schale, bevor es in das Becken gelangen konnte. In der Schale selbst liegen außerdem noch drei Photographien. Aber nur eine ist einsehbar, da die anderen beiden mit der Schichtseite nach unten gerichtet in der Schale schwimmen. Auf dieser einen Photographie ist ein wolkenverhangener Himmel über einem Meer zu sehen. Während all dies beobachtet wird, wird ein Bruchteil des im Becken befindlichen Wassers – für das menschliche Auge nicht wahrnehmbar – bereits verdunstet sein. Aufgrund einiger naturwissenschaftlicher Gesetzmäßigkeiten, die hier aber nicht weiter Erwähnung finden werden, wird es jedoch nicht verschwinden – auch wenn dies augenscheinlich der Fall sein mag – sein Aufenthaltsort wird ein anderer werden: vom Becken zur Luft. In der Luft wird es verstreut sein, so daß es als Wasser nicht mehr erkennbar ist. Aber wohin geht all das Wasser, das sich in der Luft befindet? Nur wenn es sich als Kondenswasser an den Wänden der Dunkelkammer absondern und diese herunterrinnen würde, wäre es erneut sichtbar. Da die Dunkelkammer aber über ein Belüftungssystem verfügt, ist diese Variante relativ unwahrscheinlich – die Luftfeuchtigkeit im Raum wird nicht die dafür nötige Größe erreichen. Vielmehr wird das Wasser aller Voraussicht nach durch die Belüftung – diese führt direkt auf geradem Wege vom Labor durch die Außenwand zu einer Öffnung, welche sich unterhalb der Decke befindet – nach draußen gelangen. Wird es dort, wenn es außerhalb des Labors in der Luft ist, Teil einer Wolke werden, als solche zu Regen und letztendlich wieder zu Wasser, das aus einem Wasserhahn fließt? Wird es womöglich aus genau der Wasserleitung, durch die es in das Waschbecken der Dunkelkammer kommen konnte, erneut in dieses gelangen? Theoretisch besteht sogar die, zugegebenermaßen sehr unwahrscheinliche, Möglichkeit, daß sich ein Teil des Wassers, welches momentan dabei ist zu verdunsten, bereits schon einmal genau an diesem Ort befand; es nach einer gewissen Zeit aber in das Labor zurückkehrte. Diesen Gedanken weiter verfolgt, ergibt sich die Frage, ob dieses Wasser auch Teil der Wolken, die auf der Photographie zu sehen sind, war. In diesem Falle ließe es sich in mehreren Formen an ein und demselben Ort zeitgleich betrachten. Müßig zu fragen, ob die Person, welche die Wolken photographierte und nun hier in dieser Dunkelkammer einen Abzug davon zu machen scheint, um diese Möglichkeit wußte – oder aber, ob genau dies ausschlaggebend für sie war, die Wolken zu photographieren.

Atelier
50 x 50 cm
Cibachrome on Aludibond

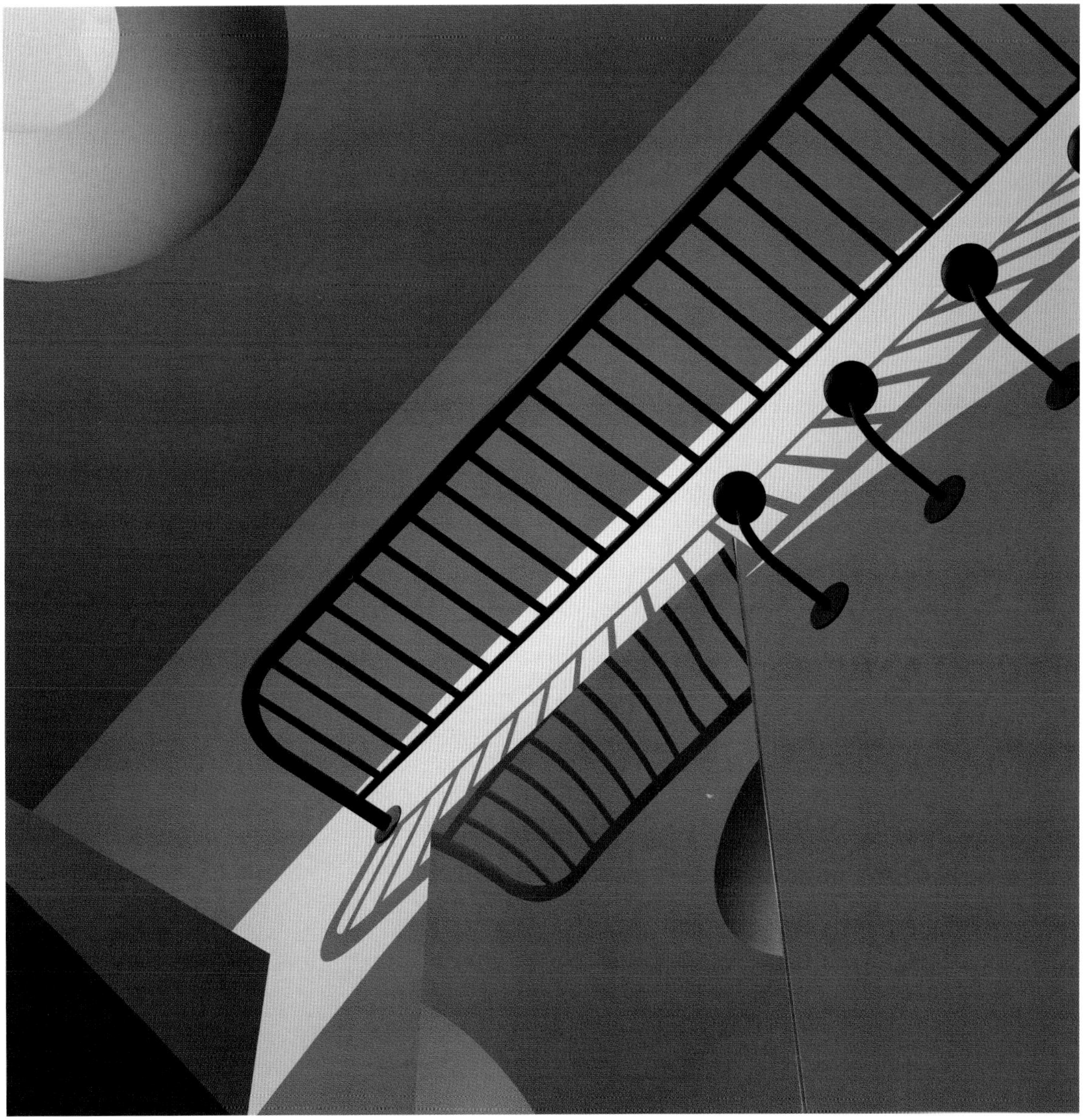

Flur
50 x 50 cm
Cibachrome on Aludibond

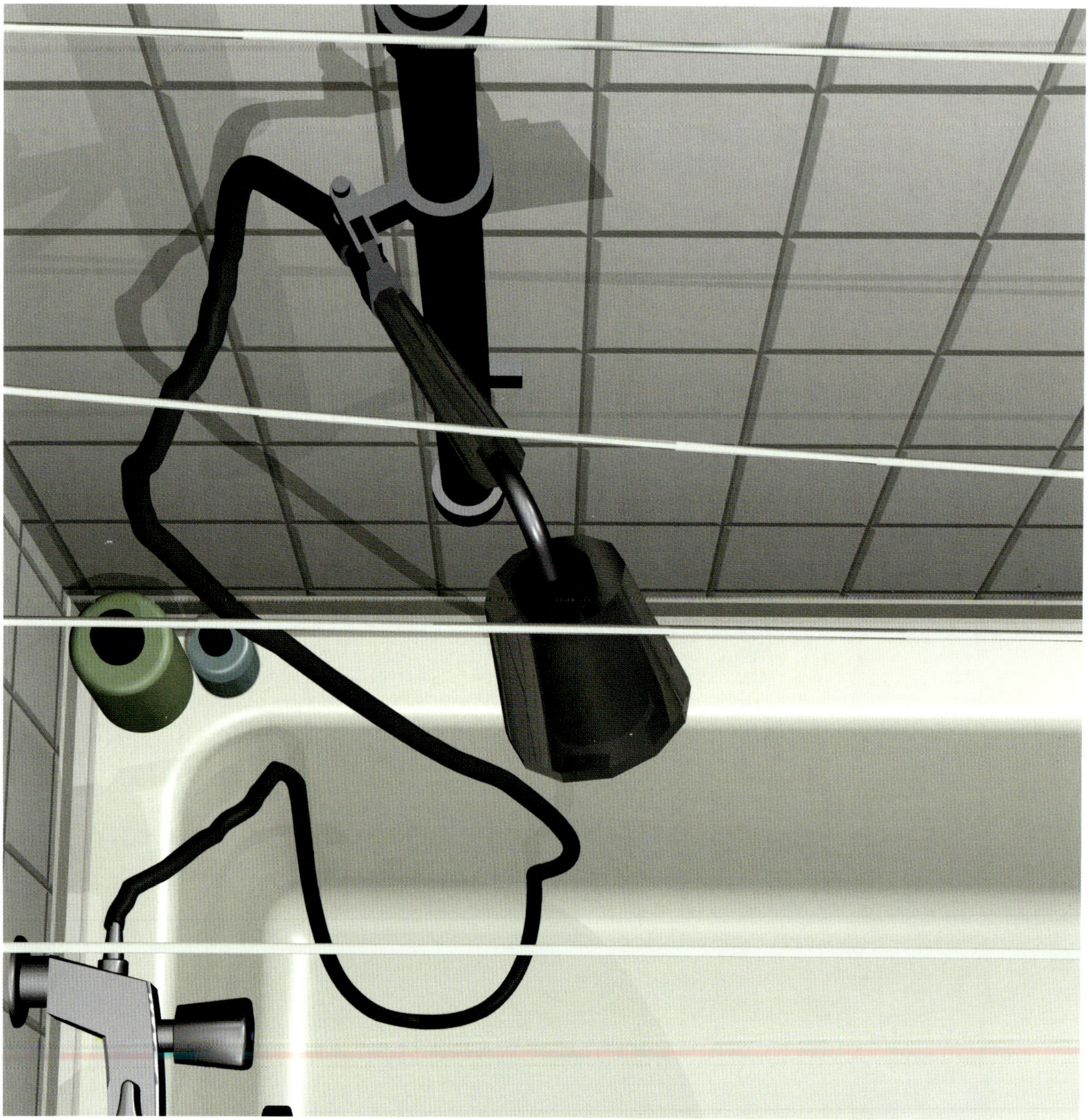

Bad I
50 x 50 cm
Cibachrome on Aludibond

Bad II
50 x 50 cm
Cibachrome on Aludibond

Diele
50 x 50 cm
Cibachrome on Aludibond

Küche
50 x 50 cm
Cibachrome on Aludibond

Küche (Detail)
50 x 50 cm
Cibachrome on Aludibond

Chinatown, New York City, 1995

Chinatown, New York City, 1995

Chinatown, New York City, 1995

Chinatown, New York City, 1995

Chinatown, New York City, 1995

Chinatown, New York City, 1995

Chinatown, New York City, 1995

The object, photographed,
Fades and then regains form.

Solid shrinks to membrane.
Membrane swells to organ.

The skin of thing
Shifts in the skin of space.

Untitled (# 204), 1997
Gelatin-silver print, 27.8 x 35.2 cm

Untitled (# 64), 1994
Gelatin-silver print, 27.6 x 30.2 cm

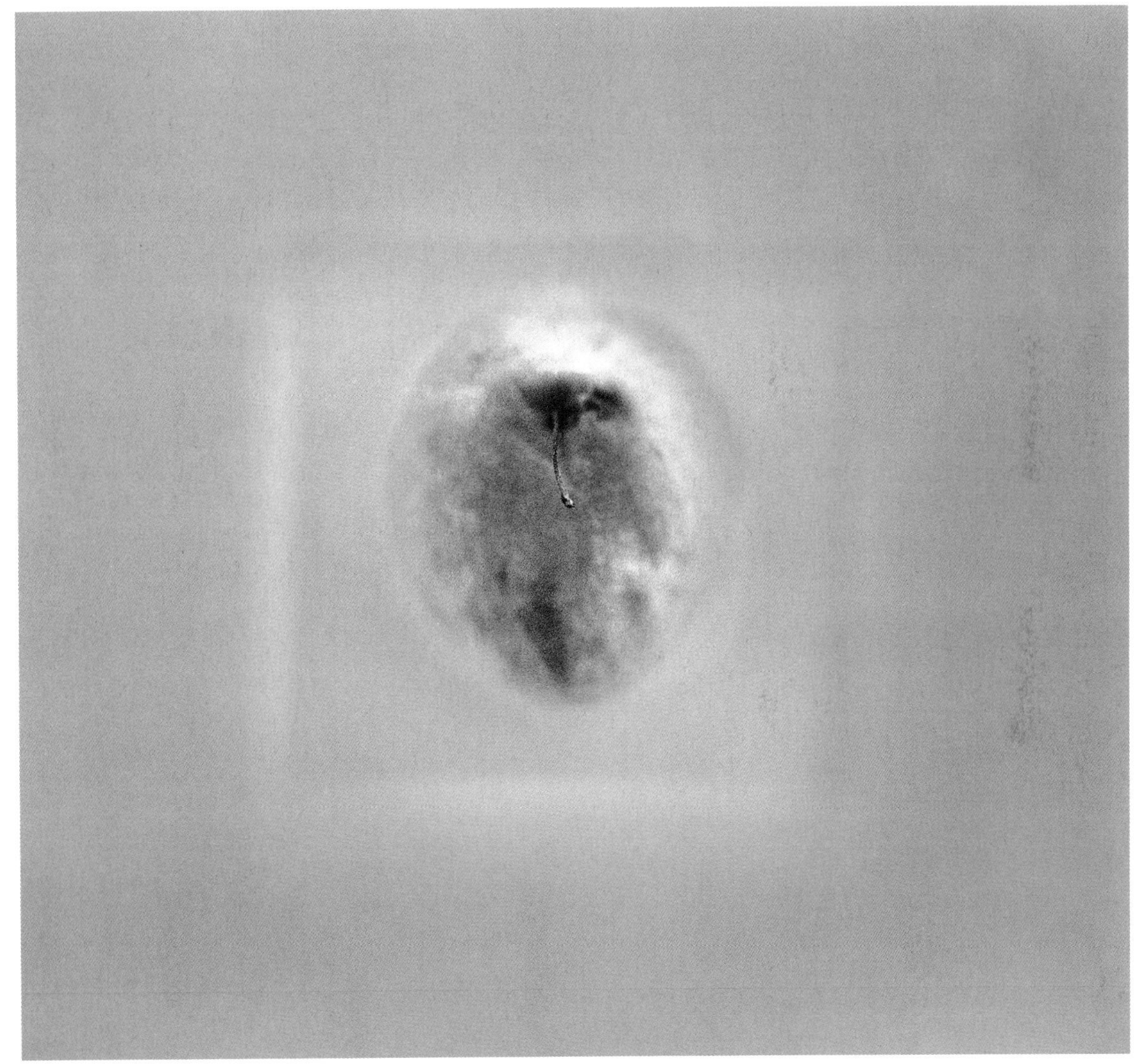

Untitled (# 175), 1997
Gelatin-silver print, 27.6 x 30.2 cm

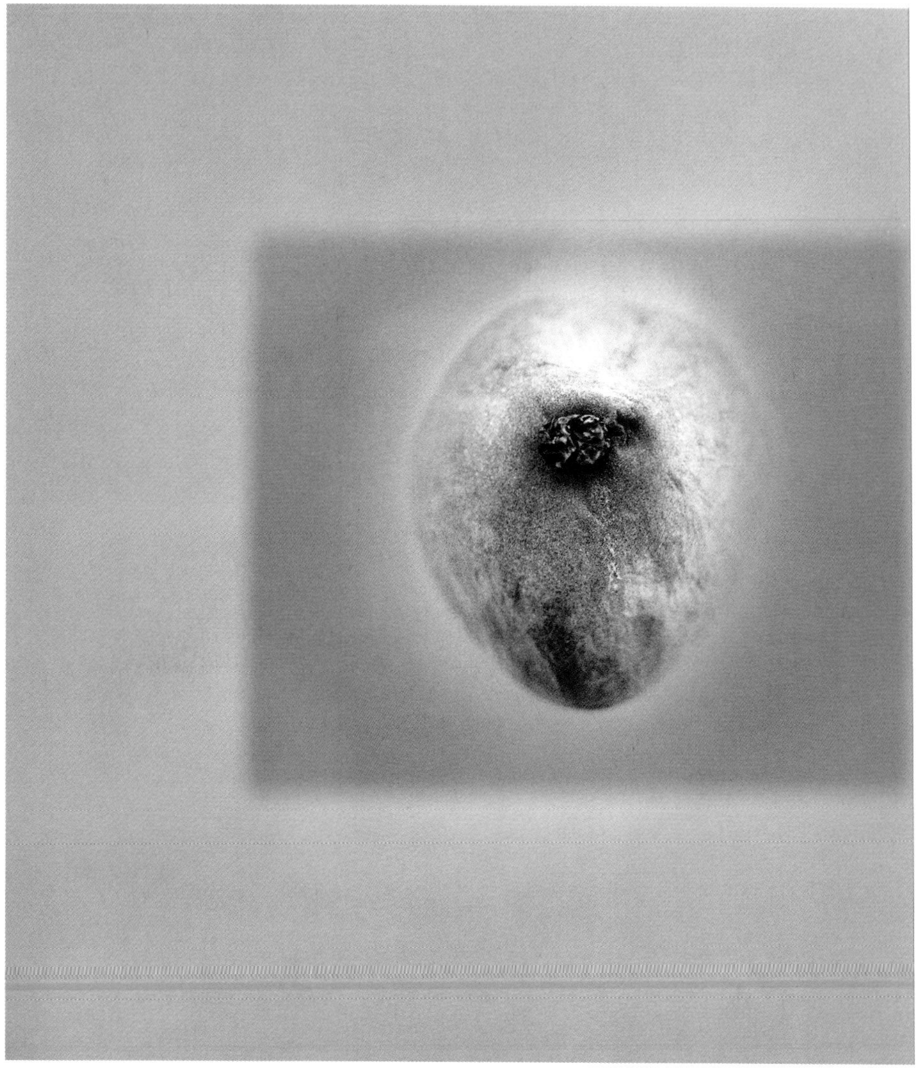

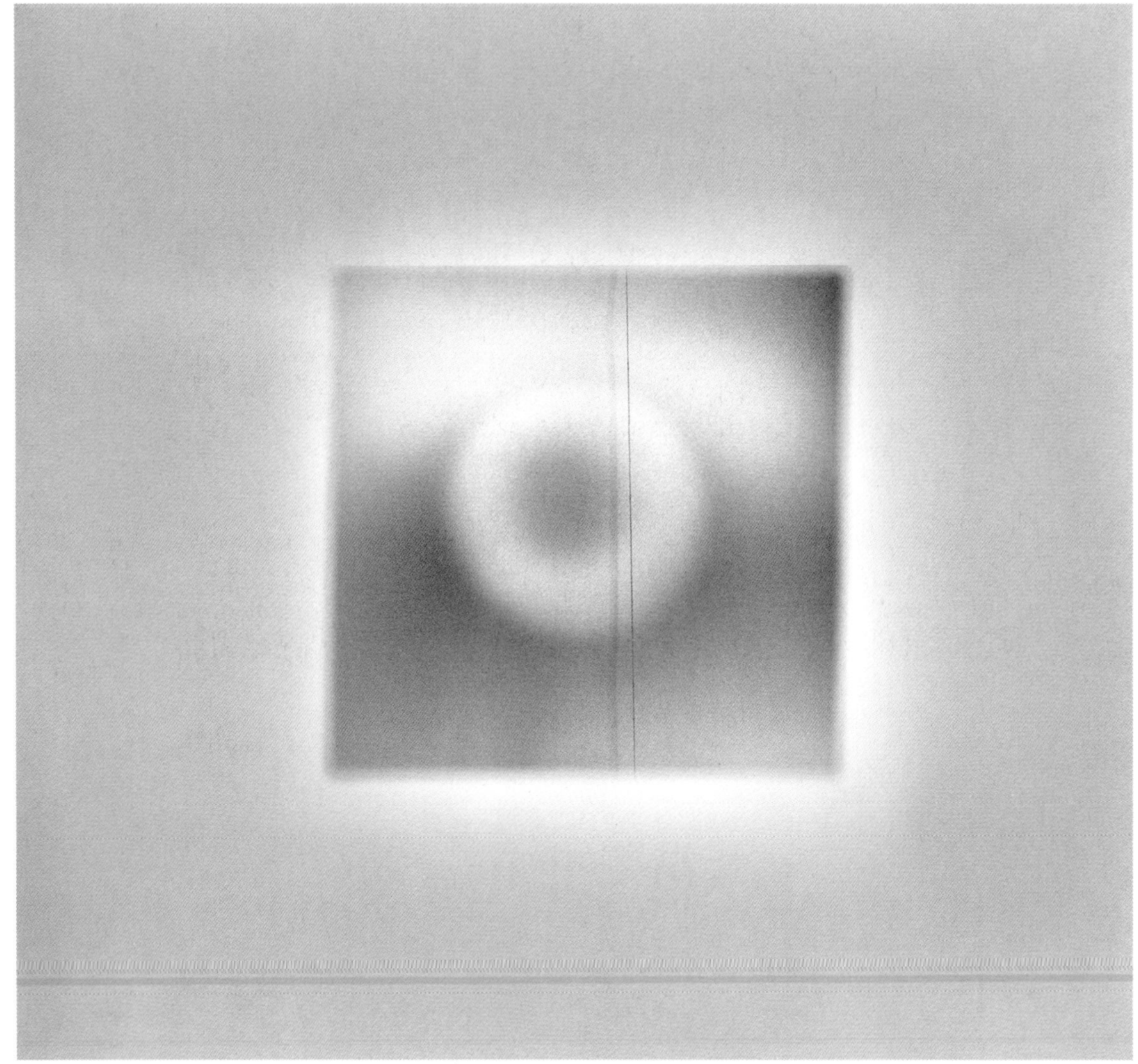

Untitled (# 63), 1992
Gelatin-silver print, 27.6 x 30.2 cm

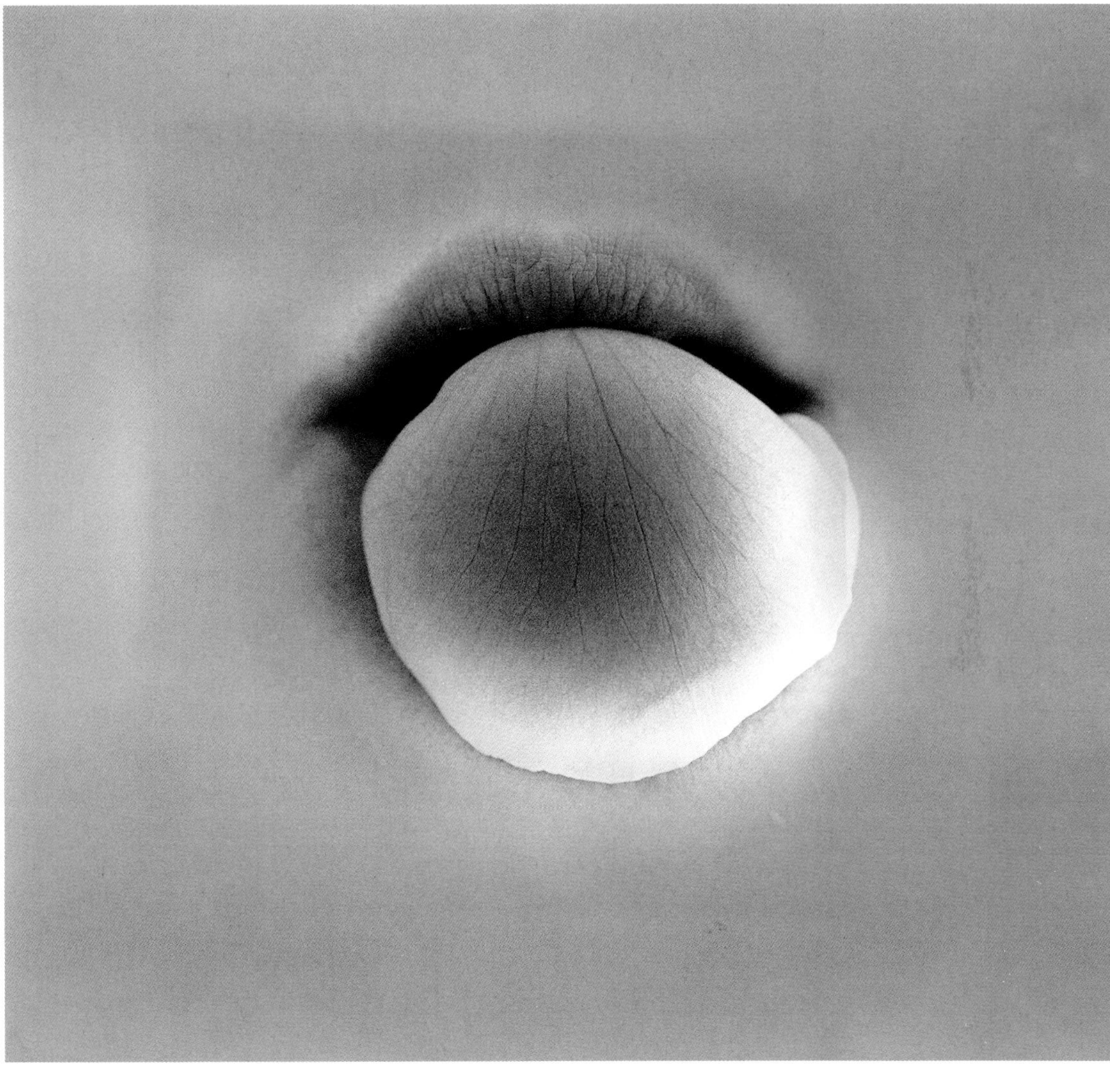

Untitled (# 120), 1995
Gelatin-silver print, 27.8 x 30.8 cm

Home and Dry II, 1996
C-print, 160 x 259 cm

A Long, Dark Night, 1994
C-print, 58 x 164 cm

Along the Bottom II, 1997
C-print, 58 x 150 cm

For my Darling, 1997
C-print, 44 x 143 cm

I remember seeing the fabulous Greek sculpture *The Kritios Boy* at the Metropolitan Museum and wondering about the time that had passed between its carving, its mutilation, the long period of burial beneath the sand like an amorphous stone and the triumph of its discovery, restoration and display in that museum. Shortly after seeing the actual piece I ran across an anonymous photograph of the same sculpture taken immediately after its exhumation from the Acropolis in 1865. Unlike the real thing, the object in the picture produced in my mind no feelings about time, memory or oblivion; no history emanated from the picture, only a sense of form, youthful form, perfect in detail and texture. Although photographs are capable of revealing information about the form and substance of the object portrayed, I was bound to see the stone surface of the young body as the skin and flesh of a live object. The sculpture no longer looked fragmented (it lacks a head and forearms) in the photograph. It looked whole and integrated into the context of its original environment.

Perhaps the photograph of *The Kritios Boy* was better than its prototype, because it could be perceived in a way more appropriate to its origin in Pericles' Greece, where the factual was often disregarded in face of the ideal. However, while the photograph narrows the spatial perception of the object, it paradoxically enhances its potential to stimulate wonder. It creates a side of the object that will always remain impenetrable to the human gaze; what lies behind of the object, its occult aspect, will be forever hidden from the viewer's eye.

The photograph does not stand for the object itself but only for a particular perspective view of it, and the best possible view is often the one that comes closest to the archetype.

During my stay in Florence in 1993 I had a chance to finally see Lorenzo Ghiberti's *Gates of Paradise.* The Renaissance masterpiece had a lasting effect on me. Here we find an exquisite command of three-point perspective combined with high relief; two anachronistic modes of representation, one numerically designed to operate within two-dimensional space and the other sculptural and analogous to three-dimensional vision. Ghiberti achieved a scale for compensation enabling one mode of representation to blend gracefully into another, creating form over a totally uncharted space.

"The hybrid or the meeting of two media is a moment of truth and revelation from which new form is born. The moment of the meeting of media is a moment of freedom and release from the ordinary trance and numbness imposed by them on our senses."
Marshall McLuhan, *Understanding Media*

To photograph a drawing is to forge a link between parallel forms of representation, to contextualize its synthetic message within the complexities of time and space.

To photograph an object is to transform it into mental substance, to map its regress into a state that predates its own existence, its return to an archetypal stage.

To photograph an archetype is to reverse the order of symbolic exchange, to retrace the trajectory of its interpretation from the image phase to the point at which it became an object in the world.

Hole, 1996
50.8 x 40.6 cm
Courtesy Camargo Vilaça, Brazil

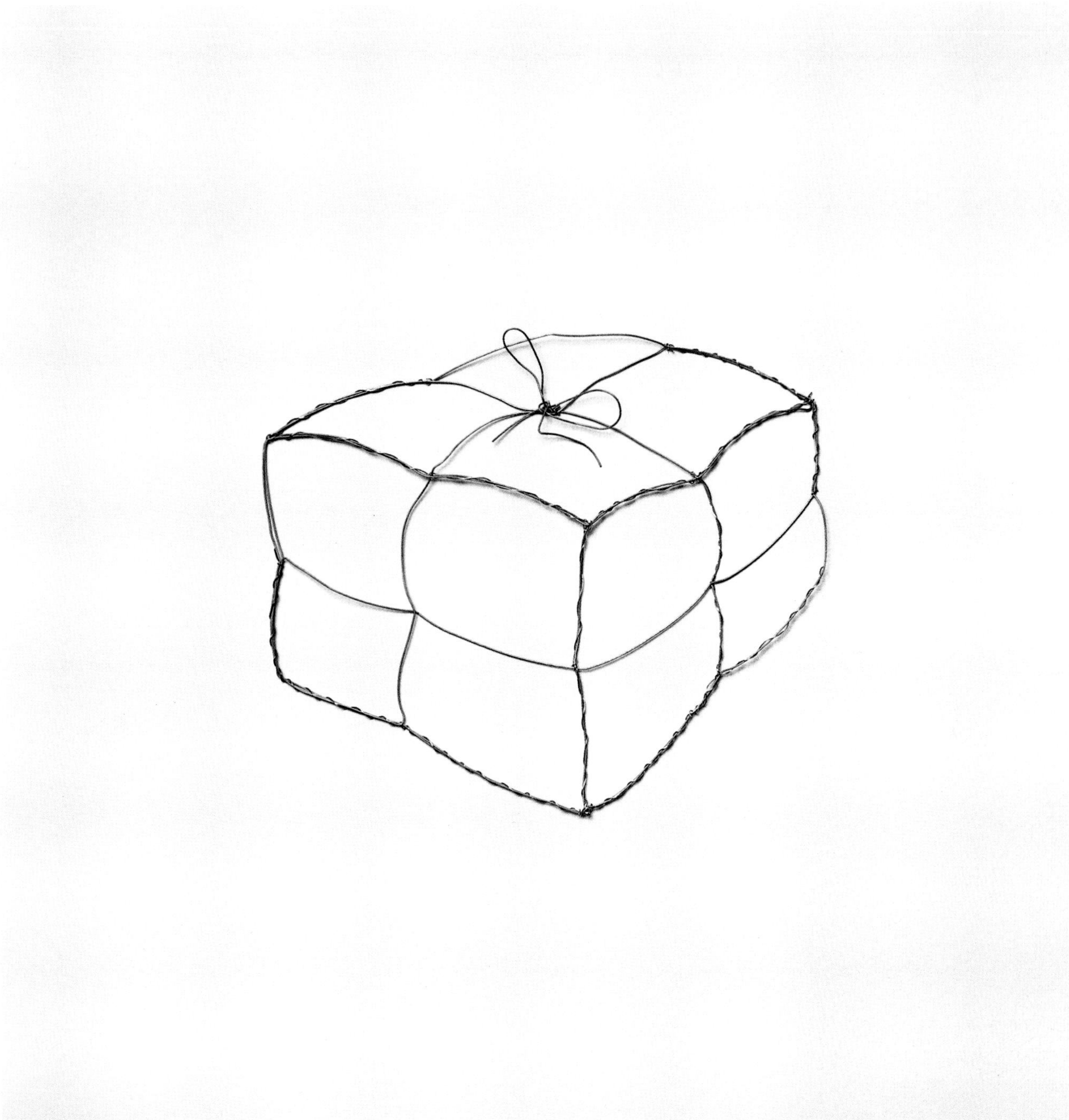

Parcel, 1995
50.8 x 40.6 cm
Courtesy Camargo Vilaça, Brazil

Relaxation, 1995
50.8 x 40.6 cm
Courtesy Camargo Vilaça, Brazil

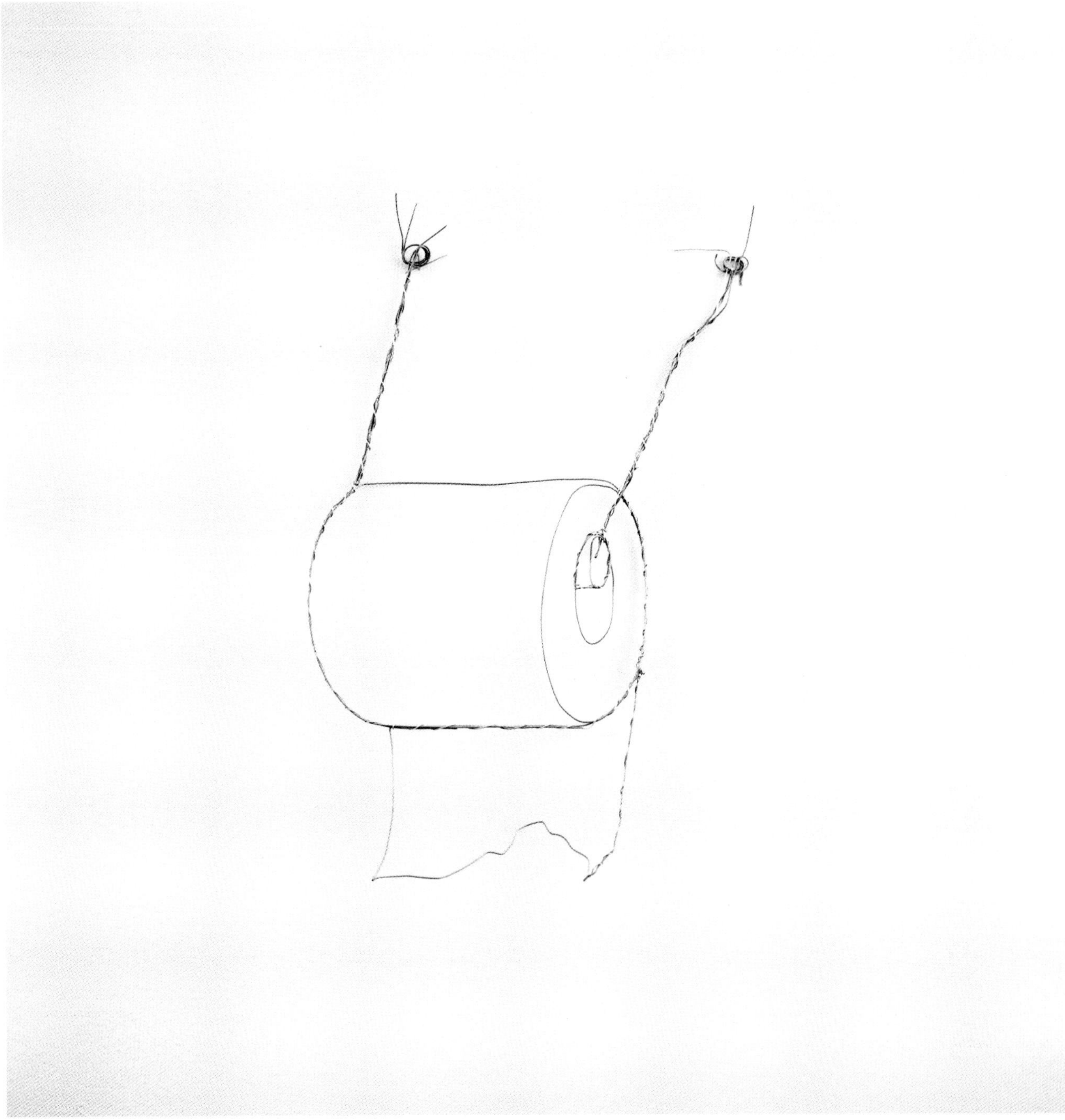

Paper & Wire, 1996
50.8 x 40.6 cm
Courtesy Camargo Vilaça, Brazil

Fiat Lux, 1995
50.8 x 40.6 cm
Courtesy Camargo Vilaça, Brazil

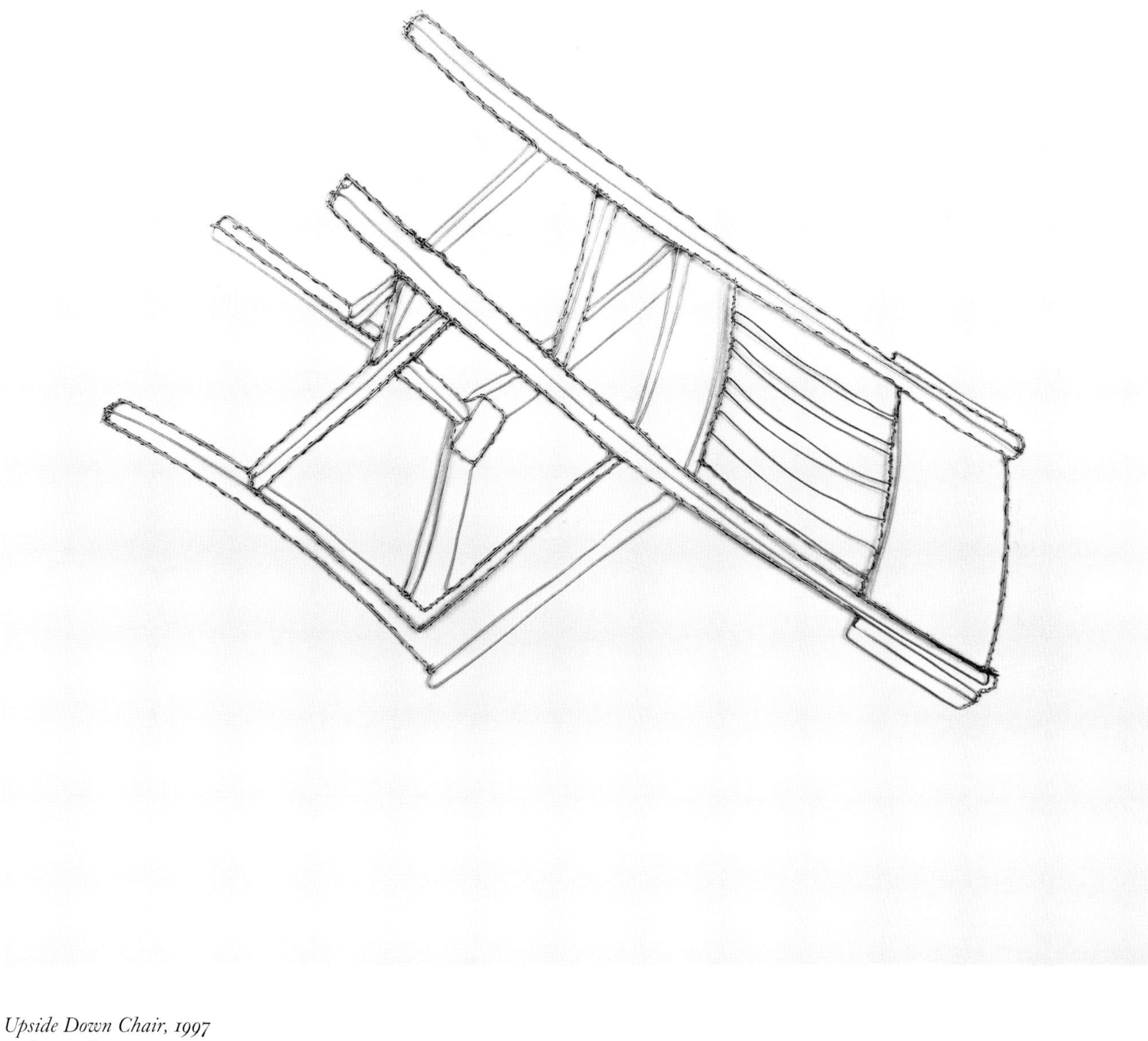

Upside Down Chair, 1997
50.8 x 40.6 cm
Courtesy Camargo Vilaça, Brazil

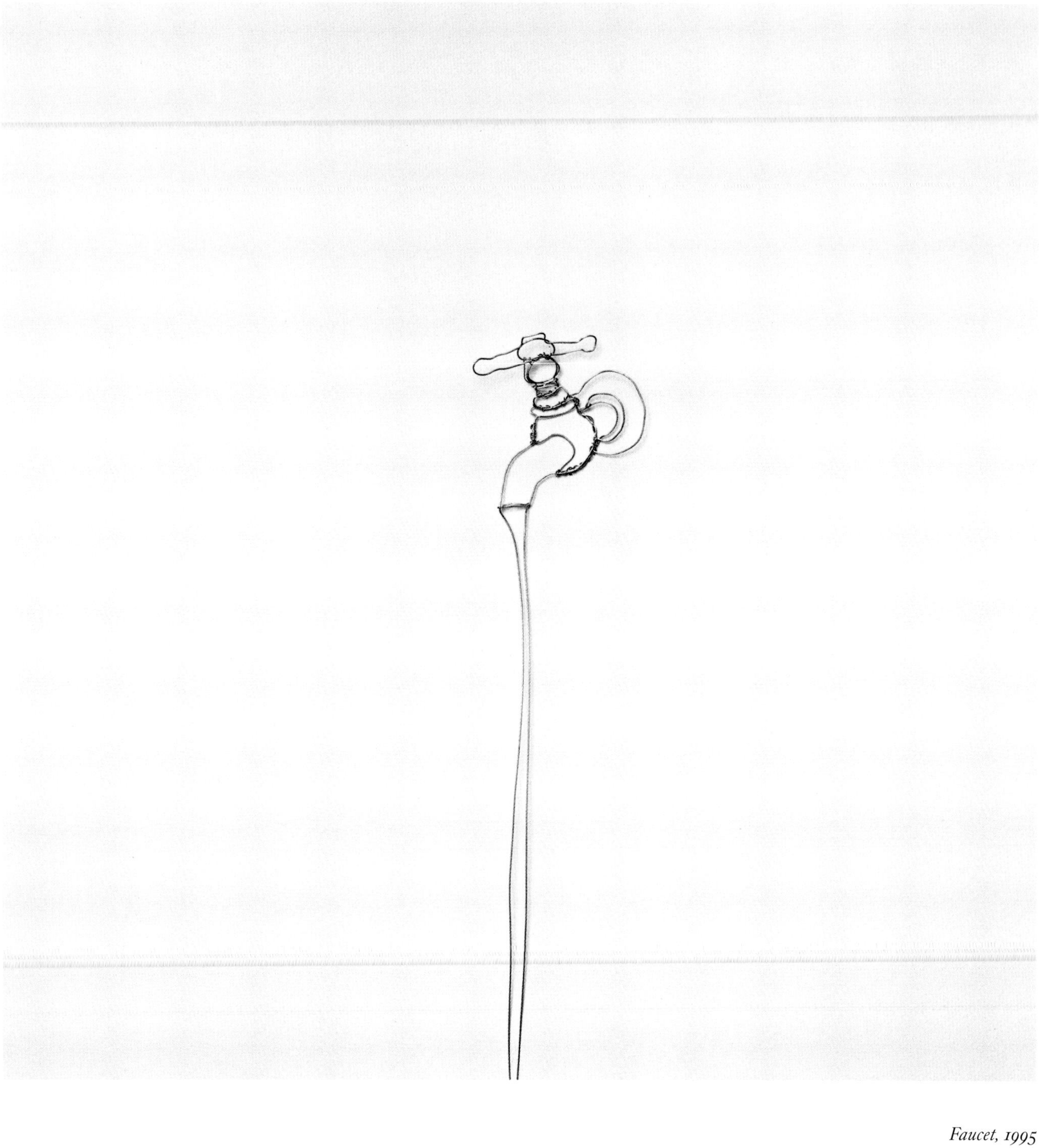

Faucet, 1995
50.8 x 40.6 cm
Courtesy Camargo Vilaça, Brazil

Sequenza 0*
Interno – Notte

Un film sull'idea di spazio molle. Un luogo umido.
La scena è interamente dipinta con un solo colore: un cosmetico
per ritoccare e correggere lo spazio.
L'illuminazione artificiale è instabile. I riflessi rendono l'ambiente vacillante
e le pareti scivolose sembrano specchi di acqua ghiacciata.
Impressione translucida. Le proporzioni cambiano continuamente:
come essere all'interno del film e poter scegliere cosa guardare.
È una percezione mobile e non definitiva. Gli oggetti sono presenze elastiche,
apparizioni impalpabili. Solo immagini trasparenti.

* Dalla sceneggiatura di un progetto per un film muto.

Tutto accadde un venerdì, 1996
Video 6'
A&M bookstore video edizioni, Milan

Esmeralda # 5, 1997
Installation, detail: slide to be projected on wall painted with interference gold acrylic and glossy iridescent varnish
Courtesy Galleria Emi Fontana, Milan

H. Cronstadt, 1997
Installation view, Chiostri di San Domenico, Imola
Color video projection 6' on wall painted
with interference gold acrylic and glossy iridescent varnish

H. Cronstadt, 1997
Installation view, Chiostri di San Domenico, Imola
Color video projection 6' on wall painted
with interference gold acrylic and glossy iridescent varnish

Le danger gluant de l'ordinaire, 1997
Installation view in the round room: 221 x 1180 cm
Musée du Papier Peint, Rixheim, France

Le danger gluant de l'ordinaire, 1997
Installation, detail: silkscreen on wall paper:
56 x 450 cm each strip, and intervention with red nail polish
Musée du Papier Peint, Rixheim, France

Tic de l'esprit, 1997
Installation View, UWM Institute of Visual Arts, Milwaukee, Wisconsin
B/W video projection 45' on wall painted with glossy iridescent varnish

1998 "Acht mal acht mal acht," Frankfurter Kunstverein, Frankfurt am Main, Germany
"Das Maß der Dinge," Ursula Blickle Stiftung, Kraichtal, Germany; Rupertinum, Galerie im Traklhaus, Salzburg, Austria

Statement

To come so close to things –
in their arrangements, surfaces and colors –
that they begin to sweat in the photograph.

Ann Mandelbaum

1945 Born in Wilkes-Barre, Pennsylvania.
Lives and works in New York City.

Adjunct Associate Professor of Photography, Pratt Institute, Brooklyn, New York City.

1963–67 Connecticut College, New London, Connecticut, Bachelor of Arts in Spanish
1976–78 New School for Social Research, New York City, Master of Arts and Sciences in Media Studies
1986–90 Pratt Institute, Brooklyn, New York City, Master of Fine Arts in Painting

Selected Solo Exhibitions
1994 Frankfurter Kunstverein, Frankfurt am Main, Germany; Espace Photographique Contretype, Brussels, Belgium; Westfälischer Kunstverein, Münster, Germany (catalogue and book)
1995 "Fotofeis," Aberdeen Art Gallery, Aberdeen, United Kingdom
Grey Art Gallery, New York City
1996 Rena Bransten Gallery, San Francisco, California
Paul Kopeikin Gallery, Los Angeles, California
1997 Galerie Françoise Paviot, Paris, France
Ubu Gallery, New York City
1998 Jackson Fine Art, Atlanta, Georgia
1998–2000 Center for Creative Photography, Tucson, Arizona; Stadtmuseum München, Fotomuseum, Munich, Germany; Stadtgalerie Saarbrücken, Germany; Musée de l'Elysée, Lausannne, Switzerland; Arge Kunst, Bolzano, Italy, Southeast Center for Contemporary Art, Wintson-Salem, North Carolina (catalogue and book)

Selected Group Exhibitions
1993 "Das Bild des Körpers," Frankfurter Kunstverein, Frankfurt am Main, Germany (catalogue and book)
1994 "La matière, l'ombre, la fiction," Bibliothèque Nationale, Paris, France (catalogue)
"Mémoire de la matière," Mai de la Photo, Reims, France (catalogue)
"The Photogram," Galerie Contretype, Brussels, Belgium; Zommer vor Photographie, Antwerp, Belgium
1995 "The Dead," National Museum of Photo, Film and T.V., Bradford, United Kingdom (traveling exhibition; catalogue)
"The Photographic Condition," Museum of Modern Art, San Francisco, California
"Inscrutable Objects," Ansel Adams Center, San Francisco, California
Vereniging Museum, Ghent, Belgium
Kölnischer Kunstverein, Cologne, Germany
1996 "The Photographic Immaterial," Canadian Cultural Center, Paris, France; Canadian Museum of Contemporary Photo, Ottawa, Canada (catalogue)
"Prospect '96," Schirn Kunsthalle Frankfurt, Frankfurt am Main, Germany (catalogue and book)
Museum of Fine Arts, Houston, Texas
1997 "The Body in the Lens," Montreal Museum of Fine Art, Montreal, Canada; Photographic Center, Skopelos, Greece
"Food, Glorious Food," Jackson Fine Art, Atlanta, Georgia
"Féminin maintenant," Galerie Françoise Paviot, Paris, France
1998 "Fictions intimes, Collection du Centre Georges Pompidou," Miró Foundation, Barcelona, Spain; Fondation Electricité de France, Paris, France
"Das Maß der Dinge," Ursula Blickle Stiftung, Kraichtal, Germany; Rupertinum, Galerie im Traklhaus, Salzburg, Austria
"Exterior/Interior – The Way I See It," Brooklyn Musem, Brooklyn, New York City

Selected Bibliography

1993 Buisine, Alain, "De-Picting the Face," *La Recherche Photographique,* Spring.
1994 *Ann Mandelbaum,* Ed. Peter Weiermair, Edition Stemmle, Kilchberg/Zürich.
Henke, Barbara, "Der Körper als Maß aller Dinge," *Handelsblatt Düsseldorf,* August.
Baer-Bogenschütz, Dorothee, "Androgyne Vortäuschungen," *Frankfurter Rundschau,* August.
Ollier, Brigitte, "Ann Mandelbaum," *La Libération,* Paris, September.
Eelbode, Erik, "Sans dire," *Bulletin Contretype,* Brussels, September.
Baque, Dominique, "Métamorphoses du désir et du dégoût," *Art Press,* December.
1995 Brittain, David, "Ann Mandelbaum," *Creative Camera,* February/March.
Coleman, A.D., "Visual Literacy – Ann Mandelbaum," *Photography in New York,* March/April.
Bonetti, David, "The Truth Isn't Black or White," *San Francisco Examiner,* San Francisco, April.
Andries, Pool, "Ann Mandelbaum," *Kunst Nu,* Ghent, May.
Coleman, A.D., "Old is New," *New York Observer,* May.
Grundberg, Andy. "Inscrutable Objects," *See* Magazine, June.
Rouillé, André, "Ann Mandelbaum," *La Recherche Photographique,* June.
McArthur, Euan, "Aberdeen Art Gallery … Ann Mandelbaum," *Portfolio* Magazine, December.
1996 "Everything That Lives Eats," *Aperture* Magazine, June.
1997 Baque, Dominique, "Mois de la photo: Itinéraires croisés," *Art Press,* January.
Langford, Martha, "The Delicate Confrontations of Ann Mandelbaum," *Border Crossings* Magazine, Summer.
Aletti, Vince, "Body Work," *The Village Voice,* New York, September.
Glueck, Grace, "Ann Mandelbaum," *The New York Times,* October.

Christopher Muller

1966 Born in Stade, Germany.
Lives and works in Düsseldorf, Germany and London, United Kingdom.

1984–85 Camberwell School of Arts & Crafts, London, United Kingdom
1985–89 Slade School of Fine Art, London, United Kingdom, Bachelor of Arts
1991–93 Audits courses under Edgar Callahan at the Staatliche Kunstakademie Düsseldorf, Germany
1995 Grant awarded by the Kunstfonds Bonn e.V.

Solo Exhibitions

1993 Räume für neue Kunst, Rolf Hengesbach, Wuppertal, Germany
1994 Städtisches Museum Abteiberg, Mönchengladbach, Germany (catalogue)
1995 Galerie Monika Reitz, Frankfurt am Main, Germany
1996 Kunstverein Freiburg, Germany (catalogue)
Forum Bildender Künstler, Essen, Germany (catalogue)
"Vom Dasein der Gegenstände," Kunstmuseum Bonn, Germany (with Richard Caldicott. (catalogue)
Galerie Monika Reitz, Frankfurt am Main, Germany
1998 Kunstverein Bochum, Germany
"Single," Räume für neue Kunst, Rolf Hengesbach, Wuppertal, Germany (with Axel Linder)

Selected Group Exhibitions

1990 Schwartzsackin Gallery, London, United Kingdom
Milch Gallery, London, United Kingdom
"Slow Horses," with AWT – Artists Working Together, London, United Kingdom
"Fünf britische Künstler," Thomas Backhaus Galerie, Düsseldorf, Germany (catalogue)
1992 "Seventeen," Greenwich Street, New York City (catalogue)
1993 "op is op," Galerie de Fabriek voor Fotografie and Goethe Institut, Rotterdam, The Netherlands; Kunstpalast Düsseldorf, Halle A1, Germany (catalogue)
1994 "Close Encounters," Ikon Gallery, Birmingham, United Kingdom
1995 Räume für neue Kunst, Rolf Hengesbach, Wuppertal, Germany
1996 "Von der Beschäftigung mit einfachen Dingen," Galerie Monika Reitz, Frankfurt am Main, Germany
"Ritratti di interno," Studio La Città, Verona, Italy
1997 "Wechselstrom," Galerie Ulrich Fiedler, Cologne, Germany
"Am Nerv der Zeit," Nassauischer Kunstverein, Wiesbaden, Germany
"Dingen," Galerie im Tor, Emmendingen, Germany
Raum III, Räume für neue Kunst, Rolf Hengesbach, Wuppertal, Germany
1998 "Heimat," Kunstverein Eislingen, Germany (catalogue)
Neue Gesellschaft für Bildende Kunst, Berlin, Germany (catalogue)
"Fotografie als Handlung," 4. Internationale Foto-Triennale Esslingen, Germany (catalogue)
"Das Maß der Dinge," Ursula Blickle Stiftung, Kraichtal, Germany; Rupertinum, Galerie im Traklhaus, Salzburg, Austria

Vik Muniz

1961 Born in São Paulo, Brazil.
Lives and works in New York City.

Solo Exhibitions

1989 Stux Gallery, New York City
1990 Stephen Wirtz Gallery, San Francisco, California
Meyers/Bloom Gallery, Santa Monica, California
1991 Galerie Claudine Papillon, Paris, France
Gabinete De Arte Rachel Arnaud, São Paulo, Brazil
Galerie Kicken-Pauseback, Cologne, Germany
1992 Claudio Botello Arte, Turin, Italy
"Individuals," Stux Gallery, New York City
1993 "Equivalents," Galleria Ponte Pietra, Verona, Italy; Tricia Collins Contemporary Art, New York City (catalogue)
1994 "Representations," Wooster Gardens, New York City
1995 "The Wire Pictures," Galeria Camargo Vilaça, São Paulo, Brazil
1996 "Pantomimes," Rena Bransten Gallery, San Francisco, California
"The Best of Life," Wooster Gardens, New York City
Galeria Casa da Imagen, Curitiba, Brazil
"The Sugar Children," Tricia Collins Contemporary Art, New York City
1997 Dan Bernier Gallery, Los Angeles, California
Galeria Camargo Vilaça, São Paulo, Brazil
"Pictures of Thread," Wooster Gardens, New York City
1998 Rena Bransten Gallery, San Francisco, California
"Seeing is Believing," International Center of Photography, New York City
Galeria Módulo, Lisbon, Portugal
Galeri Lars Bohman, Stockholm, Sweden

Selected Group Exhibitions

1989 "De Rozeboomkamer," Beeldenroute Foundation, Diepenheim, The Netherlands
"Wortlaut: Konzepte zwischen visueller Poesie & Fluxus," Galerie Schuppenhauer, Cologne, Germany
1990 "Non Sculpture," Galerie Barbara Farber, Amsterdam, The Netherlands
"Constructed Illusion," Pace MacGill Gallery, New York City
"Assembled," Wright State University, Dayton, Ohio
"On the Edge, Between Sculpture and Photography," Cleveland Center for Contemporary Art, Cleveland, Ohio
1991 "Mike Kelley/Vik Muniz/Jim Shaw," Real Art Ways, Hartford, Connecticut
"The Encompassing Eye, Photography as Drawing," University Art Galleries, University of Akron, Ohio (traveled)
"The Neighborhood," A.I.R., Amsterdam, The Netherlands
"Anni Novanta," Galleria d'Arte Moderna, Bologna, Italy
"The Fetish of Knowledge," Real Art Ways, Hartford, Connecticut
1992 "Life Size: Small, Medium, Large," Museo D'Arte Contemporanea Luigi Pecci, Prato, Italy
"Multiples," The Aldrich Museum of Art, Ridgefield, Connecticut
"Theoretically Yours," Regione Autonoma della Valle d'Aosta, Aosta, Italy
"Detour," The International House, New York City (catalogue)
1993 "The Alternative Eye: Photography for the 90's," Southern Alleghenies Museum, Loreto, Pennsylvania
"Time to Time," Castello de Rivara, Turin, Italy
"Sound," Museo d'Arte Moderna, Bolzano, Italy
"Post-Verbum," Palazzo della Regione Bergamo, Bergamo, Italy
1994 "Single Cell Creatures," Katonah Museum of Art, Katonah, New York City

"Crash," The Thread Waxing Space, New York City
"Up the Establishment: Reconstructing the Counterculture," Sonnabend Gallery, New York City
"Jetlag," Galerie Martina Detterer, Frankfurt am Main, Germany

1995 "Changing Perspectives," Contemporary Art Museum, Houston, Texas
"Panorama Da Arte Contemporânea Brasileira, Museu de Arte Morderna, São Paulo, Brazil (catalogue)
"Mostra America," Fundação cultural de Curitiba, Curitiba, Brazil
"The Photographic Condition," San Francisco Museum of Modern Art, San Francisco, California

1996 "Inclusion/Exclusion" Steirischer Herbst, Graz, Austria
"Some Assembly Required," The Art Institute of Chicago, Chicago, Illinois
"Shadow Play," San Jose Institute of Contemporary Art, San Jose, California

1997 "New Photography 13," The Museum of Modern Art, New York City
"Assi esta la cosa," Centro Cultural Arte Contemporáneo, Mexico D.F., Mexico
"20 Years, Almost," Robert Miller Gallery, New York City
"Hope," The National Arts Club, New York City

1998 XXIV Bienal Internacional de São Paulo, Brazil
"Antropofagia Urbana," Paço das Artes, São Paulo, Brazil
"The Garden of the Forking Paths," Kunstforeningen, Copenhagen, Denmark; Oslo, Norway; Helsinki, Finland
"Das Maß der Dinge," Ursula Blickle Stiftung, Kraichtal, Germany; Rupertinum, Galerie im Traklhaus, Salzburg, Austria

Selected Bibliography

1990 Bonetti, David, "The Inevitability of Vik Muniz," *San Francisco Examiner,* April 6.
Baker, Kenneth, "Eccentric and Entertaining Pieces," *The San Francisco Chronicle,* April 7.
Aletti, Vince, "Constructed Illusions," *The Village Voice,* New York City, November 20.

1991 Levin, Kim, "Choices," *The Village Voice,* New York City, April 2.

1992 Morgan, Stuart, "Vik Muniz at Prato," *Frieze Magazine,* No. 7.
Hagen, Charles, "The Encompassing Eye; Photography as Drawing," *Aperture,* No. 125, New York City, Fall.
Cotter, Holland, "Three Shows Celebrate the Spirit of Fluxus," *The New York Times,* October 23.

1993 Hagen, Charles, "Vik Muniz," *The New York Times,* February 5.
Levin, Kim, "Vik Muniz" *In Brief/The Village Voice,* New York City, February 16.

1995 Aletti, Vince, "Choices," *The Village Voice,* New York City, March 21.
Hagen, Charles, "Mixing Humor with the History of Photography," *The New York Times,* March 24.
Katz, Vincent, "Vik Muniz at Brent Sikkema," *Art in America,* December, p. 98.

1996 Bonetti, David, "Vik Muniz at Rena Bransten," *San Francisco Examiner,* May 10.
Aletti, Vince, "Choices," *The Village Voice,* New York City, December 3.
Smith, Roberta, "Art in Review," *The New York Times,* December 13.
Mahoney, Robert, "Vik Muniz, Sugar Children," *Time Out New York,* December 19.

1997 Cotter, Holland, "200 Images Leisurely Chatting," *The New York Times,* January 31.
Kandel, Susan, "Artful Illusions and Serial Masquerades Spring to Life," *Los Angeles Times,* April 4.
Grundberg, Andy, "Sweet Illusion," *Artforum,* September.
Kismarik, Susan, "Vik Muniz, New Photography 13," *Museum of Modern Art Magazine,* New York City, Fall.
Cotter, Holland, "New Photography 13," *The New York Times,* November 7.
Aletti, Vince, "Organized Confusion," *The Village Voice,* New York City, December 2.

1998 Goldberg, Vicki, "Of Fairies, Free Spirits and Outright Frauds," *The New York Times,* February 1, Weekend Section, Front Page.
Miller, Kenneth, "Sugar Snaps," *Life Magazine,* March.
Leffingwell, Edward, "Through a Brazilian Lens," *Art in America,* April, p. 78.
Baker, Kenneth, "Muniz Whets the Appetite for Images in Chocolate," *The San Francisco Chronicle,* April 17.

Peter Peryer

1941 Born in Auckland, New Zealand.
Lives and works in Auckland.

Selected Solo Exhibitions

1977 "Peter Peryer: An Introduction," Dowse Art Museum, Lower Hutt, New Zealand

1985 "Peter Peryer/Photographs," Sarjeant Gallery, Wanganui, New Zealand (traveled New Zealand; catalogue)

1992 "Recent Work," Govett Brewster Art Gallery, New Plymouth, New Zealand (traveled New Zealand)

1995 "Second Nature," Frankfurter Kunstverein, Frankfurt am Main; Ludwig Forum, Aachen, Germany (traveled New Zealand and Australia; catalogue and book)

1997 "At Home and Away," Sarjeant Gallery, Wanganui, New Zealand (catalogue)

1997–98 Esbjerg Kunstmuseum, Esbjerg, Denmark; Arken Museum, Arken, Denmark

Selected Group Exhibitions

1975 "The Active Eye," Manawatu Art Gallery, Palmerston North, New Zealand (traveled New Zealand)

1979 "Three New Zealand Photographers," Auckland City Art Gallery, Auckland, New Zealand (traveled New Zealand; catalogue)

1982 "Views/Exposures: Ten contemporary New Zealand photographers," National Art Gallery, Wellington, New Zealand (traveled New Zealand)
"Vision in Disbelief: The 4th Biennale of Sydney," Art Gallery of New South Wales, Sydney, Australia (catalogue)

1983 "International Photography 1980–82," National Gallery of Australia, Canberra, Australia (traveled Australia)

1984 "Anxious Images: Aspects of recent New Zealand art," Auckland City Art Gallery, Auckland, New Zealand (traveled)

1986 "Content/Context: A survey of recent New Zealand art," Shed 11, National Art Gallery, Wellington, New Zealand

1989 "From Today Painting is Dead," Shed 11, National Art Gallery, Wellington, New Zealand
"Imposing Narratives: Beyond the documentary in recent New Zealand Photography," Wellington City Art Gallery, Wellington, New Zealand (traveled New Zealand; catalogue)

1990 “Two Centuries of New Zealand Landscape Art,” Auckland City Art Gallery, Auckland, New Zealand

1991 “175° East,” Sarjeant Gallery, Wanganui, New Zealand (traveled Australia)

1994 “Pictograms,” Monash University Gallery, Victoria, Australia (traveled Australia)

1996 10th Biennale of Sydney, New Zealand representative
“Beyond Recognition,” National Gallery of Australia (traveled Australia)

1997 “The Enigmatic Object,” Art Gallery of New South Wales, Sydney, Australia

1998 “The Dream Collectors,” Museum of New Zealand, Te Papa, New Zealand
“Eat!” Museum of Contemporary Art, Sydney, Australia
“Das Maß der Dinge,” Ursula Blickle Stiftung, Kraichtal, Germany; Rupertinum, Galerie im Traklhaus, Salzburg, Austria

Awards

1997 New Zealand Order Of Merit, bestowed by Queen Elizabeth II.

Selected Bibliography

1975 “Peter Peryer,” *Photo-Forum 29,* Wellington, December, pp. 10–15 (portfolio with statement).

1976 “Peter Peryer,” *Photo-Forum 33,* Wellington, August/September, pp. 14–17 (portfolio with statement).

1977 Fraser, Ross, “Peter Peryer: The photograph as a portrait of the self,” *Art New Zealand 8,* November/December, pp. 25, 65, 67.

1977/78 Barr, Jim, “Peter Peryer,” *Photo-Forum Supplement 1,* Wellington, Summer, pp. 9–15 (interview).

1978 “Peter Peryer,” *Creative Camera International Yearbook,* London, pp. 113–123.

1979 “Three New Zealand Photographers,” *Photo-Forum Supplement 3,* Wellington, Spring, p. 4.
Turner, John, “ New Zealand Photography since 1945,” *Printletter, 23,* No. 4/5, September/October, Zurich, pp. 14–18.

1981 “A desire to understand,” *New Zealand Listener,* February 7, pp. 36–37.
Keith, Sheridan, “Frontierland – The Photographs of Peter Peryer,” *London Magazine 21,* July 4, pp. 42–49.

1982 Curnow, Wystan, “New Zealand art – where in the world is it?,” *New Zealand Listener,* April 3, pp. 34–35.

1984 Ireland, Peter, “The Significance of Repetition,” *Art New Zealand 32,* pp. *34–35.*

1985 Barr, Jim and Barr, Mary, *Peter Peryer,* Sarjeant Art Gallery, Wanganui.

1989–90 Elias, Ann, "Peter Peryer," *Art New Zealand 53,* Summer, p. 80.

1991 Eldredge, Charles C., *Pacific Parallels,* San Diego Museum of Art, San Diego, pp. 47–48, 50.
Headlands – Thinking through New Zewland art, Museum of Contemporary Art, Sydney, pp. 46, 72.

1992 Wedde, Ian, “A Shark in Time,” *Midwest 1,* Govett Brewster Art Gallery, New Plymouth, p. 19.

1995 *Second Nature, Peter Peryer, Photographer, New Zewland,* Edition Stemmle, Kilchberg/Zurich.

Alessandra Tesi

1969 Born in Bologna, Italy.
Lives and works in Bologna, Italy and Paris, France.

1994 Diploma at Accademia di Belle Arti, Bologna, Italy

1995 Institut des Hautes Études en Arts Plastiques, Paris, France, Directeur Fondateur Pontus Hulten

Selected Solo Exhibitions

1995 Spazio Cultura Navile, Bologna, Italy

1996 Castello di Rivara, Italy (catalogue)
Galleria Paolo Vitolo, Milan, Italy (catalogue)

1997 Spazio Aperto, Galleria d’Arte Moderna, Bologna, Italy (catalogue)
“Le danger gluant de l’ordinaire,” Musée du Papier Peint, Rixheim, France
“Tic de l’esprit,” Galleria Neon, Bologna, Italy
“Rosso D-R,” Institute of Visual Arts, University Art Museum of Wisconsin, Milwaukee

1998 “Voyager,” Santa Maria della Scala, Siena, Italy

Selected Group Exhibitions

1994 “Equinozio d’autunno 1994,” Castello di Rivara, Italy
“We are moving,” Viafarini, Milan, Italy

1995 “Out of order, Aperto ’95,” Galleria d’Arte Moderna, Bologna, Italy (catalogue)
“Il giovane ospite/Neon,” Casa di Giorgione, Castelfranco Veneto, Italy

1996 “Prospect ’96,” Schirn Kunsthalle Frankfurt, Frankfurt am Main, Germany (catalogue and book)
“Carte Italiane,” Art Center Eleftherias Park, Athens, Greece (catalogue)
“Spazi Comuni,” Galerie Esca, Nîmes, France
“Il luogo e la memoria,” Ex-Copertificio Sonnino, Varese, Italy; Castello di Spezzano, Italy (catalogue and book)
“Il Punto,” Galleria Continua, San Gimignano, Italy
“Panorama Italiano 1,” Trevi Flash Art Museum, Trevi, Italy (catalogue)
“XII Quadriennale Nazionale,” Palazzo delle Esposizioni, Stazione Termini, Rome, Italy (catalogue)
“Jahresgaben 1996,” Frankfurter Kunstverein, Frankfurt am Main, Germany

1997 “Giro d’Italia,” Galleria L’Attico, Rome, Italy
Galerie Martina Detterer, Frankfurt am Main, Germany
“Camera Preview,” Art & Public, Geneva, Switzerland
“504,” Zentrum für Kunst, Medien und Design, Braunschweig, Germany
“Modi e luoghi,” La Posteria, Mailand, Italy, (catalogue)
“Des histoires en formes,” Le Magasin, Centre National d’Art Contemporain, Grenoble, France
“Portraits of Interiors,” Studio La Città, Verona, Italy; Galerie Blancpain Stepcynski, Geneva, Scwitzerland; Patricia Faure Gallery, Los Angeles, California (catalogue)
“Vertigo,” Galleria Emi Fontana, Milan, Italy
“Officina Italia,” Bologna; Imola; Castel S. Pietro; Cesena; Forlì; Santa Sofia, Italy (catalogue)
“[illegible] Man World,” Internet project (http://www.undo.net/beyond)
“Fatto in Italia, films and videos,” Centre d’Art Contemporain, Geneva, Switzerland; Institute of Contemporary Arts, London, United Kingdom (catalogue)

1998 "Habitat," Associazione Culturale Valentina Moncada, Rome, Italy
"Subway," subway stations, Milan, Italy
"Das Maß der Dinge," Ursula Blickle Stiftung, Kraichtal, Germany; Rupertinum, Galerie im Traklhaus, Salzburg, Austria
"Opera Nuova / Fuori Uso," Ex-Mercato Ortofrutticolo, Pescara, Italy
"Liberamente," Spazio Ex-Arrigoni, Cesena, Italy (catalogue)
"Côté Sud … Entschuldigung," Frac Rhône-Alpes, Le Nouveau Musée, Villeurbanne, France

Selected Bibliography

1995 Tesi, Alessandra, "Ouverture," *Flash Art,* No. 193.
1996 Pinto, Roberto, *Spazi Comuni,* Papiers Libres, Nîmes.
Christov-Bakargiev, Carolyn, "Avanguardia 96," *L'Espresso,* No. 11.
Marziani, Gianluca, "Silenzi pulsanti di Alessandra Tesi," *L'Opinione,* No. 42.
Curto, Guido, "Alessandra Tesi," *Flash Art,* No. 198.
1997 Barilli, Renato, "Nati da un'officina," *L'Espresso,* No. 11.
Cerizza, Luca, "Alessandra Tesi," *Juliet,* No. 83.
Hartmann, Serge, "Alessandra Tesi au Musée du Papier Peint," *DNA Alsace,* June 2.
Cereja, Pierre-Louis, "Tesi," *L'Alsace,* June 6.
Daolio, Roberto, "Alessandra Tesi," *Flash Art,* No. 205.
Pagel, David, "Paintings in Midst of Generational Conflict," *Los Angeles Times,* August 1.
Vettese, Angela, "Cercando il gioco in una realtà mutante," *Il Sole 24 Ore,* October 12.
De Cecco, Emanuela, "Vertigo," *Flash Art,* No. 207; *Flash Art International,* No. 198.
Colasanti, Claudia, "Officina Italia," *Flash Art,* No. 207.
1998 Jacqué, Bernard/Pajano, Rosalba/Tesi, Alessandra, "Cosmesi dello spazio," *ars mediterranea,* Barcelona.
Jacqué, Bernard/Pajano, Rosalba/Pasini, Francesca/Tesi, Alessandra, *Hygiène,* Le Quai Editions, Mulhouse.

Statement

Scene 0*
Indoors – Night

A film on the concept of soft space. A moist place. The scene is immersed in a single color – a cosmetic used to retouch and correct the space. The artificial light is erratic. Reflections make the surroundings appear to flicker, and the slippery walls have the look of mirrors of ice. They seem translucent. Proportions are constantly changing – as if we were inside the film and could choose what we would like to see. An alterable, indefinite kind of perception. The objects are elastic presences, manifestations impossible to grasp. Mere transparent images.

* From the script for a silent film project

This book is published in conjunction with the exhibition entitled "Das Maß der Dinge," presented at the Ursula Blickle Foundation, Kraichtal from June 7th to July 5th, 1998, and the Galerie im Traklhaus, Salzburg from September 24th to October 31st, 1998.

The exhibition curator and publication editor for the Ursula Blickle Foundation is Peter Weiermair.

© 1998 by Ursula Blickle Stiftung, Kraichtal, Germany and
EDITION STEMMLE AG, Thalwil/Zurich, Switzerland

All rights reserved.
No part of the book may be reproduced or transmitted in any form or by any means, electronic or mechanical, including photocopy, recording or any other information storage and retrieval system, nor may any part of the book be reprinted in newspapers or periodicals, used in public addresses, films or dramatizations or broadcast on radio or television without the written permission of the publishers. This applies as well to individual illustrations and text passages.

Works copyright by the artists
Texts copyright by the authors

Translations from the German texts by John S. Southard
Editorial direction by Esther Oehrli
Art direction by renntypo, visuelle kommunikation, Teufen, Switzerland
Printed and bound by Kündig Druck AG, Baar, Switzerland

ISBN 3-908161-48-7